POLAR BEARS
AT THE ZOO

written by Mia Coulton
photographed by Mia Coulton & Amy Musser

Here comes the biggest **bear** at the zoo.

It is the **polar** **bear.**

Polar bears have two small ears, two small eyes and a big black nose.

Polar bears can smell things that are very far away.

The polar bear has black skin underneath its **fur**.

The black skin **absorbs** the sun's heat and helps the polar bear stay warm.

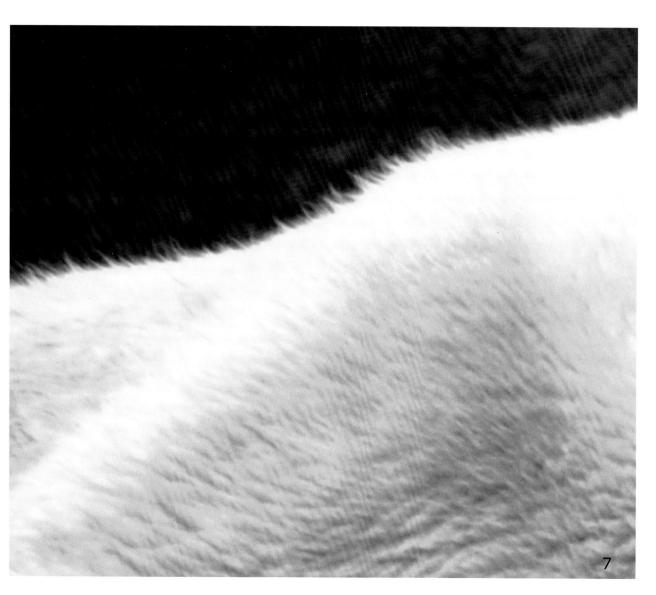

7

Look at the polar bear's **paw**. It is huge!

The big paws of the polar bear have **footpads** like **Velcro**. This keeps the polar bear from slipping on the ice.

Look at the polar bear swimming. Its front paws have **webbed** toes that help it swim.

Polar bears are good swimmers.

Here is a baby polar bear.

A baby polar bear is called a **cub**.

This cub was born at the zoo.

Polar bears are fun to watch at the zoo.

Glossary

absorbs: Taking in or soaking up

bear: A medium to large animal with heavy fur, long claws and a very short tail; types include polar bears, black bears and brown bears

cub: A baby bear

footpads: The thick, spongy skin on the bottom of the toes and feet of most furry animals

fur: Hair covering an animal's body

paw: The foot of a furry animal, with pads and claws

polar bear: A large white bear that lives near the North Pole

Velcro: Popular brand of hook-and-loop binding, inspired by the prickly burrs found in woods that stick to clothing and fur

webbed: Describes fingers or toes that are connected by a thin fold of skin